ENERGY SECTOR STANDARD
OF THE PEOPLE'S REPUBLIC OF CHINA

中华人民共和国能源行业标准

Specification for Geological Observation
of Hydropower Projects

水电工程地质观测规程

NB/T 35039-2014

Chief Development Department: China Renewable Energy Engineering Institute

Approval Department: National Energy Administration of the People's Republic of China

Implementation Date: March 1, 2015

China Water & Power Press

中国水利水电出版社

Beijing 2024

All rights reserved. No part of this publication may be reproduced, stored in a retrieval system, or transmitted in any form or by any means—electronic, mechanical, photocopying, recording or otherwise, without prior written permission of the publisher.

图书在版编目（CIP）数据

水电工程地质观测规程：NB/T 35039-2014 = Specification for Geological Observation of Hydropower Projects (NB/T 35039-2014) : 英文 / 国家能源局发布. -- 北京：中国水利水电出版社，2024.3. -- ISBN 978-7-5226-2805-9
Ⅰ. P642-65
中国国家版本馆CIP数据核字第2024LP9972号

ENERGY SECTOR STANDARD
OF THE PEOPLE'S REPUBLIC OF CHINA
中华人民共和国能源行业标准

Specification for Geological Observation
of Hydropower Projects
水电工程地质观测规程
NB/T 35039-2014
（英文版）

Issued by National Energy Administration of the People's Republic of China
国家能源局　发布
Translation organized by China Renewable Energy Engineering Institute
水电水利规划设计总院　组织翻译
Published by China Water & Power Press
中国水利水电出版社　出版发行
　　Tel: (+ 86 10) 68545888　68545874
　　sales@mwr.gov.cn
　　Account name: China Water & Power Press
　　Address: No.1, Yuyuantan Nanlu, Haidian District, Beijing 100038, China
　　http://www.waterpub.com.cn
中国水利水电出版社微机排版中心　排版
北京中献拓方科技发展有限公司　印刷
184mm×260mm　16开本　4.5印张　142千字
2024年3月第1版　2024年3月第1次印刷
Price（定价）：￥740.00

Introduction

This English version is one of China's energy sector standard series in English. Its translation was organized by China Renewable Energy Engineering Institute authorized by National Energy Administration of the People's Republic of China in compliance with relevant procedures and stipulations. This English version was issued by National Energy Administration of the People's Republic of China in Announcement [2021] No. 1 dated January 7, 2021.

This version was translated from the Chinese Standard NB/T 35039-2014, *Specification for Geological Observation of Hydropower Projects*, published by China Electric Power Press. The copyright is reserved by National Energy Administration of the People's Republic of China. In the event of any discrepancy in the implementation, the Chinese version shall prevail.

Many thanks go to the staff from the relevant standard development organizations and those who have provided generous assistance in the translation and review process.

For further improvement of the English version, any comments and suggestions are welcome and should be addressed to:

China Renewable Energy Engineering Institute
No. 2 Beixiaojie, Liupukang, Xicheng District, Beijing 100120, China
Website: www.creei.cn

Translating organization:

POWERCHINA Huadong Engineering Corporation Limited

Translating staff:

    PENG Peng   NI Weida   ZHOU Lipei   ZHANG Qifeng

Review panel members:

LIU Xiaofen	POWERCHINA Zhongnan Engineering Corporation Limited
GUO Jie	POWERCHINA Beijing Engineering Corporation Limited
LI Weichao	China Institute of Water Resources and Hydropower Research
QIAO Peng	POWERCHINA Northwest Engineering Corporation Limited

QIE Chunsheng	Senior English Translator
QI Wen	POWERCHINA Beijing Engineering Corporation Limited
CHEN Li	POWERCHINA Kunming Engineering Corporation Limited
ZHAO Yanrong	Hohai University
YIN Desheng	China Three Gorges University
ZHOU Zhou	POWERCHINA Guiyang Engineering Corporation Limited
HUANG Chen	POWERCHINA Chengdu Engineering Corporation Limited
LIU Aimei	Beijing Millennium Engineering Technology Corporation Limited
LI Zhongjie	POWERCHINA Northwest Engineering Corporation Limited
WANG Huiming	China Renewable Energy Engineering Institute

National Energy Administration of the People's Republic of China

翻译出版说明

本译本为国家能源局委托水电水利规划设计总院按照有关程序和规定，统一组织翻译的能源行业标准英文版系列译本之一。2021年1月7日，国家能源局以2021年第1号公告予以公布。

本译本是根据中国电力出版社出版的《水电工程地质观测规程》NB/T 35039—2014 翻译的，著作权归国家能源局所有。在使用过程中，如出现异议，以中文版为准。

本译本在翻译和审核过程中，本标准编制单位及编制组有关成员给予了积极协助。

为不断提高本译本的质量，欢迎使用者提出意见和建议，并反馈给水电水利规划设计总院。

地址：北京市西城区六铺炕北小街2号
邮编：100120
网址：www.creei.cn

本译本翻译单位：中国电建集团华东勘测设计研究院有限公司
本译本翻译人员：彭　鹏　倪卫达　周力沛　章奇锋
本译本审核人员：

刘小芬　中国电建集团中南勘测设计研究院有限公司

郭　洁　中国电建集团北京勘测设计研究院有限公司

李维朝　中国水利水电科学研究院

乔　鹏　中国电建集团西北勘测设计研究院有限公司

郄春生　英语高级翻译

齐　文　中国电建集团北京勘测设计研究院有限公司

陈　砺　中国电建集团昆明勘测设计研究院有限公司

赵燕容　河海大学

殷德胜　三峡大学

周　洲　中国电建集团贵阳勘测设计研究院有限公司

黄　晨　中国电建集团成都勘测设计研究院有限公司

刘爱梅　北京木联能工程科技有限公司

李仲杰　中国电建集团西北勘测设计研究院有限公司
王惠明　水电水利规划设计总院

国家能源局

Announcement of National Energy Administration of the People's Republic of China [2014] No. 11

According to the requirements of Document GNJKJ [2009] No. 52, "Notice on Releasing the Energy Sector Standardization Administration Regulations (*tentative*) and detailed implementation rules issued by National Energy Administration of the People's Republic of China", 330 sector standards such as *Carbon Steel and Low Alloy Steel for Pressurized Water Reactor Nuclear Power Plants—Part 17: Pushed Formed Elbows for Use in Main Steam Systems*, including 71 energy standards (NB), 122 electric power standards (DL) and 137 petroleum and natural gas standards (SY), are issued by National Energy Administration of the People's Republic of China after due review and approval.

Attachment: Directory of Sector Standards

National Energy Administration of the People's Republic of China

October 15, 2014

Attachment:

Directory of Sector Standards

Serial number	Standard No.	Title	Replaced standard No.	Adopted international standard No.	Approval date	Implementation date
...						
48	NB/T 35039-2014	Specification for Geological Observation of Hydropower Projects			2014-10-15	2015-03-01
...						

Foreword

According to the requirements of Document FGBGY [2008] No. 1242 issued by General Office of National Development and Reform Commission of the People's Republic of China, "Notice on Releasing the Development Plan of the Sector Standards in 2008", and after extensive investigation and research, summarization of practical experience, consultation of relevant Chinese and foreign advanced standards and wide solicitation of opinions, the drafting group has prepared this specification.

The main technical contents of this specification include: fault and seismic activity observation, reservoir-induced earthquake observation, groundwater observation, slope deformation observation, unstable rock mass observation, debris flow observation, geotechnical observation on foundations, and surrounding rock observation in underground cavern.

National Energy Administration of the People's Republic of China is in charge of the administration of this specification. Powerchina Renewable Eergy Engineering Institute has proposed this specification and is responsible for its routine management. Energy Sector Standardization Technical Committee on Hydropower Investigation and Design is responsible for the explanation of specific technical contents. Comments and suggestions in the implementation of this specification should be addressed to:

China Renewable Energy Engineering Institute
No. 2 Beixiaojie, Liupukang, Xicheng District, Beijing 100120, China

Chief development organization:

POWERCHINA Huadong Engineering Corporation Limited

Participating development organizations:

Zhejiang Huadong Construction Engineering Corporation Limited

Zhejiang Huadong Engineering Safety Technology Corporation Limited

Huadong Engineering Corporation Fujian Regional Headquarters

Zhejiang Huadong Surveying, Mapping and Geoinformation Corporation Limited

Chief drafting staff:

SHAN Zhigang	LU Fei	SHI Anchi	MENG Yongxu
LIU Shiming	YE Zhiping	ZHOU Huixin	ZHOU Chunhong
SHI Jianwei	NIE Jinsheng		

Review panel members:

PENG Tubiao	LI Shisheng	ZHU Jianye	GUO Yihua
HUANG Xiaohui	FAN Junxi	ZHOU Zhifang	CHEN Zhijian
LIU Aimei	GONG Hailing	YANG Tianjun	XIA Hongliang
CUI Changwu	ZHANG Guofu	ZHANG Sihe	

Contents

1	**General Provisions**	1
2	**Terms**	2
3	**Basic Requirements**	4
4	**Fault and Seismic Activity Observation**	6
4.1	General Requirements	6
4.2	Fault Activity Observation	6
4.3	Seismic Activity Observation	8
5	**Reservoir-Induced Earthquake Observation**	12
5.1	General Requirements	12
5.2	Arrangement of Observation Networks	12
5.3	Technical Requirements of Observation System	13
5.4	Observation and Operation	14
5.5	Data Collation and Analysis	14
6	**Groundwater Observation**	15
6.1	General Requirements	15
6.2	Observation Contents and Range	15
6.3	Observation Arrangement	15
6.4	Observation Methods	17
6.5	Observation Frequency and Accuracy	20
6.6	Data Collation and Analysis	22
7	**Slope Deformation Observation**	24
7.1	General Requirements	24
7.2	Observation Items and Contents	24
7.3	Observation Arrangement	24
7.4	Observation Methods	26
7.5	Observation Frequency and Accuracy	29
7.6	Data Collation and Analysis	30
8	**Unstable Rock Mass Observation**	32
8.1	General Requirements	32
8.2	Observation Items and Contents	32
8.3	Observation Arrangement	32
8.4	Observation Methods	33
8.5	Data Collation and Analysis	34
9	**Debris Flow Observation**	35
9.1	General Requirements	35
9.2	Observation Items and Contents	35
9.3	Observation Arrangement	36

9.4	Observation Methods	37
9.5	Data Collation and Analysis	38
10	**Geotechnical Observation on Foundations**	**39**
10.1	General Requirements	39
10.2	Observation Items and Methods	39
10.3	Observation Frequency and Data Processing	40
11	**Surrounding Rock Observation in Underground Cavern**	**42**
11.1	General Requirements	42
11.2	Observation Contents and Methods	42
11.3	Observation Frequency and Data Collation	43
Appendix A	**Engineering Geological Observation Items of Each Investigation Stage**	**45**
Appendix B	**Comprehensive Classification of Active Faults**	**46**
Appendix C	**Scales and Main Types of Unstable Rock Mass**	**48**
Appendix D	**Classification of Debris Flow and Erosion Degree**	**50**
Appendix E	**Deformation Observation Items of Underground Cavern and Tunnel Surrounding Rock**	**57**
Explanation of Wording in this Specification		**59**
List of Quoted Standards		**60**

1 General Provisions

1.0.1 This specification is formulated with a view to unifying the contents, methods and technical requirements of geological observation for hydropower projects to ensure the quality of observation results.

1.0.2 This specification is applicable to the observation work in geological investigation for large- and medium-sized hydropower projects (including pumped-storage power stations).

1.0.3 In addition to this specification, the geological observation of hydropower projects shall comply with other current relevant standards of China.

2 Terms

2.0.1 slope

natural or cut slope composed of the crest, surface, and toe, including potential unstable, deformation, and landslide bodies

2.0.2 unstable rock mass

potential collapse mass on a slope, mainly including unstable rock block, loose rock mass, slipping-tension rock mass, toppling rock mass, etc.

2.0.3 debris flow

mixture of loose rock, soil, water and gas flowing along a natural slope or gully

2.0.4 rockburst

phenomenon that rock cracks abruptly, accompanied with loud sound or rock shoot during excavation caused by rapid release of a large amount of elastic strain energy which is accumulated in deep-seated hard rock mass under high ground stress

2.0.5 harmful gas

gas or steam in an underground cavern that is harmful to human and ecological environment, mainly including CO, H_2S, SO_2, NH_3, CH_4, CO_2, H_2, nitrogen oxide, etc.

2.0.6 ground temperature

general term for rock and soil temperature at different depths

2.0.7 observation frequency

times of observation over a certain period

2.0.8 observation period

time interval between two consecutive observations

2.0.9 observation line of longitudinal direction

observation line approximately parallel to the direction of slope deformation

2.0.10 observation line of transverse direction

observation line approximately perpendicular to the direction of slope deformation

2.0.11 trend value

value at the center point of a data set, which reflects the extent of a set of data being close to the central value

2.0.12 deformation rate

relative deformation amount of slope or unstable rock mass per unit time, which reflects the average deformation over a certain period, including daily, weekly, monthly, and yearly deformation rates

2.0.13 standard sheet erosion

soil erosion on a soil slope other than gully erosion, which is caused by runoff when the rainfall intensity exceeds infiltration capacity of ground, thin-sheet flow, or micro-stream flow, carrying away the fine particles of the topsoil, and leaving micro or small grooves and scaly pits on the slope surface

2.0.14 erosion degree

extent of denudation damage to topsoil by the action of natural force (water, wind, gravity, freeze-thaw, etc.) and human activities

3 Basic Requirements

3.0.1 The engineering geological observation is intended to provide basic data, analysis and predictions for identifying the main engineering geological conditions and evaluating the main engineering geological problems in the project area.

3.0.2 The engineering geological observation items include fault and seismic activity, reservoir-induced earthquake, groundwater, slope, unstable rock mass, debris flow, soil and rock in foundation, surrounding rock in underground caverns, etc.

3.0.3 The engineering geological observation shall be carried out according to the needs of engineering geological investigation in a project area, and shall be arranged and analyzed commensurate with the work requirements of the corresponding investigation stage. The observation items for each investigation stage shall comply with Appendix A of this specification.

3.0.4 The engineering geological observation shall be arranged on the basis of the collection and analysis of the available data on hydrometeorology, topography and geology, and field reconnaissance in the project area, and the engineering geological observation outline should be prepared.

3.0.5 An engineering geological observation outline shall include:

1 A general description of engineering geological conditions and available engineering geological investigation results.

2 The purpose, assignment and working conditions.

3 The contents, methods and technical requirements.

4 Workload and schedule.

5 Observation equipment and personnel.

6 Safety and quality assurance measures.

7 Observation layout.

3.0.6 Engineering geological observation shall meet the following requirements:

1 Appropriate observation method and equipment shall be selected according to the importance and hazard of the observed object, topographic and geologic conditions, and technical feasibility and economy. The observation equipment shall be inspected and calibrated as required.

2   The layout of observation points shall make full use of existing boreholes, pits, exploratory adits, shafts and other exploration work.

3   The observation shall be improved continuously as the investigation goes deeper. Any abnormal results shall be analyzed and reported timely, and the frequency of observation shall be increased accordingly.

4   The observation data shall be processed and analyzed timely.

3.0.7   At the end of each stage of the engineering geological observation work, an engineering geological observation analysis report shall be developed, which shall present the conclusion, trend prediction and suggestion, and be annexed with necessary drawings and attachments.

4 Fault and Seismic Activity Observation

4.1 General Requirements

4.1.1 Fault activity observation shall be conducted if a project encounters the following conditions:

1 Active faults exist within 5 km of the project site.

2 A large (Grade 1) project with a dam height over 200 m or reservoir storage capacity more than 10 billion m^3, or a large (Grade 1) project with a dam height over 150 m in the region with a basic seismic intensity of Ⅶ and above.

4.1.2 Seismicity observation shall be conducted if a project encounters the following conditions:

1 Active faults exist within 5 km of the project site; the geological structure is complex, seismic activity is frequent, and the observation ability of the professional seismic network is weak.

2 A large (Grade 1) project with a dam height over 200 m or reservoir storage capacity more than 10 billion m^3, or a large (Grade 1) project with a dam height over 150 m in the region with a basic seismic intensity of Ⅶ and above.

4.1.3 The observation of faults and seismic activities should be started from the pre-feasibility study stage.

4.2 Fault Activity Observation

4.2.1 The fault activity observation shall include deformation or displacement.

4.2.2 The observation methods may include cross-fault short leveling, cross-fault short baseline, cross-fault ranging and triangulation network, Global Positioning System (GPS) network, connecting tube, extensometer, inclinometer, etc. If conditions permit, multiple methods should be used for each observation point. All kinds of observation shall meet the following requirements:

1 The vertical deformation of fault shall be observed according to the first-order leveling accuracy, and the specific methods shall comply with the current national standard GB/T 12897, *Specifications for the First and Second Order Leveling*. For an area with steep terrain, trigonometric leveling measurements may be adopted, and the specific

methods shall comply with the current sector standard DB/T 47, *The Method of Earthquake-Related Crust Monitoring—Fault-Crossing Displacement Measurement*.

2   For horizontal deformation of fault, the observation methods of GPS network, short baseline and triangulation network shall be adopted, and these methods shall comply with the current standards of China GB/T 18314, *Specifications for Global Positioning System (GPS) Surveys*; and DB/T 47, *The Method of Earthquake-Related Crust Monitoring—Fault-Crossing Displacement Measurement*.

3   The observation period of fault displacement shall be once every two months. The observation frequency may be increased as appropriate when the fault deformation is significantly beyond the normal value. The observation duration shall not be less than two years.

4.2.3   The arrangement of observation points shall meet the following requirements:

1   The observation stations/points may be arranged across the fault in one to several representative sections selected along the main fault zone. The places where the observation stations are arranged shall have clear fault exposure, obvious activity sign, suitable topography for using various observation methods, and shall be convenient for access and communication.

2   The observation points crossing the fault shall be arranged according to the requirements of the observation method, which shall form a reasonable combination, and the points must be arranged on stable rock mass of both sides beyond the fault-influenced zone. In river bank slope with steep terrain, the observation points of fault deformation shall be arranged beyond the sections influenced by rock mass gravity deformation.

3   In addition to the above requirements, the point selection for cross-fault short leveling, GPS and cross-fault triangulation network shall comply with the current standards of China GB/T 12897, *Specifications for the First and Second Order Leveling*; GB/T 18314, *Specifications for Global Positioning System (GPS) Surveys*; and DB/T 47, *The Method of Earthquake-Related Crust Monitoring—Fault-Crossing Displacement Measurement*.

4   The fault deformation observed by the cross-fault short baseline, the water pipe inclinometer and the extensometer should be carried out in

the cavern crossing the fault. The observation line may be set up to be vertical to the fault or to form an intersection angle of about 30°, and the relatively constant temperature and less environmental disturbance are required in the cavern.

4.2.4 The observation data processing and result submitting shall meet the following requirements:

1. Prepare the station/point sheets as well as the report and drawings on the construction of deformation monitoring network according to relevant requirements.

2. Check the deformation observation data obtained by various methods, and compile the results such as duration curves, vector diagrams and corresponding tables, respectively.

3. Prepare the fault activity observation report annually. If sudden change of fault deformation occurs in a certain period within one year, a brief report shall be prepared in time and submitted to relevant department. The comprehensive classification of activity fault may be in accordance with Appendix B of this specification.

4. Present the observation results after the observation work or phased observation work is finished, including the observation report, corresponding drawings, tables, and other related attachments, and compile the reports and drawings into volumes for filing by period.

4.3 Seismic Activity Observation

4.3.1 The observation range should be within 20 km to 40 km of the project area. For an extremely large or important project, the observation range may be expanded appropriately, which should cover the earthquake controlling fault nearby.

4.3.2 The seismic activity observation content shall include occurring time, spatial location, intensity, frequency, etc.

4.3.3 The arrangement of the observation network shall meet the following requirements:

1. For the areas near the project area where geological structure is complex and seismic activity is frequent, single or mobile station should be set up in pre-feasibility study stage in order to understand the characteristics of seismic activity.

2. When there is high probability of earthquake with a magnitude above

Ⅵ in the project site, a special seismic observation network shall be set up for hydropower projects.

3 Before the observation network is set up, an overall design report shall be prepared. The observation station/network shall be well set up so that the seismic activity in the vicinity of the main structure can be observed.

4 The observation network shall consist of at least 4 seismic stations and the digital telemetry network shall be used whenever possible. The distribution of the seismic stations shall be uniform and reasonable and form multiple triangles to meet the requirements for effective observation accuracy.

4.3.4 The site selection of the seismic station shall comply with the current sector standard DB/T 16, *Specification for the Construction of Seismic Station Seismograph Station*, especially the following requirements:

1 The station shall be located outside the prospective project construction area and the planning construction area, and shall avoid arterial roads, woods, rivers, mines and all kinds of potential interference sources. The distance away from the interference sources and the requirements of the observation stations for noise level of the observational environment shall comply with the current national standard GB/T 19531.1, *Technical Requirement for the Observational Environment of Seismic Stations—Part 1: Seismometry*.

2 The selected station site shall be set on intact and hard bedrock with a large area of outcrop, and shall be free of fault zone. The station should not be established on loose deposits formation and weathered layer, otherwise special foundation treatment shall be conducted.

3 The location of the station should be flat, and the steep slope, wind gap and geological disasters affected areas shall be avoided. The attended station site should be selected in the relatively low area. For a telemetry station, the channel requirements shall be met firstly.

4 The station site should be provided with electricity, communication and traffic conditions, and the attended station shall also be provided with living and working conditions.

5 For selection of a station site, trial observation shall be conducted for one month. Only after every indicator meets the requirements of station site selection can the location of each station can be determined. And the station site selection report and corresponding charts shall be

submitted.

4.3.5 Earthquake observation methods and requirements shall meet the following requirements:

1. The digital telemetry network shall be adopted for observation whenever practical.

2. For attended stations, instruments shall be calibrated regularly, the working status of instruments checked when needed, real-time waveform display reviewed timely, and data processed and analyzed preliminarily. The data should be regularly backed up and the duty log recorded.

3. For the recorded microearthquake, identification shall be conducted for whether the microearthquake is induced by blasting vibration or other human activities.

4. During seismic observation, when any felt earthquake with magnitude $M_L \geq 2.5$ near the main structures, or that with magnitude $M_s \geq 5$ within 150 km of the main structures occurs, the owner of the project needs to be informed within 15 min after the earthquake occurrence.

Translator's Annotation: This item is translated in accordance with Announcement [2019] No. 6, Attachment 3.4—Sector Standard Amendment Notification issued by National Energy Administration of the People's Republic of China on November 4, 2019.

5. When an earthquake of $M_L \geq 3.5$ occurs within the coverage of the observation network, site investigation shall be organized to determine the macroscopic epicenter and isoseismal distribution.

4.3.6 The processing of seismic observation data and submitting of results shall meet the following requirements:

1. Analyze and process in time all the earthquake data recorded by the network, and establish seismic records.

2. Check and compile digital seismic records, magnetic media records or drawings of each station, and establish a seismic database.

3. Prepare an annual seismic catalogue, epicenter distribution map, and seismic network observation report for each observation year.

4. Compile ten-day or monthly seismic activity report based on seismic activity, and report to the project owner and designer, submit daily or weekly report according to actual situation in case of abnormality. Input

the compiled seismic data into the database in time.

5   Propose a site seismic survey report attached with seismo-geological map, isoseismal map, photos, video for the earthquake of $M_L \geq 3.5$ within observation range.

5 Reservoir-Induced Earthquake Observation

5.1 General Requirements

5.1.1 Special forecast and evaluation of reservoir-induced earthquake shall be conducted for a large-scale reservoir in the pre-feasibility study and feasibility study stages. Reservoir-induced earthquake observation shall be carried out for newly-built reservoirs of which the dam is higher than 100 m and the reservoir storage capacity is more than 500 million m^3. Necessary earthquake monitoring facilities shall be set up to closely monitor the seismic activities for the newly-built reservoirs, of which active faults pass through the highest water level reserve and its 10 km expansion area that serious secondary disasters may occur after earthquake damage.

Translator's Annotation: This article is translated in accordance with Announcement [2019] No. 6, Attachment 3.4—Sector Standard Amendment Notification issued by National Energy Administration of the People's Republic of China on November 4, 2019.

5.1.2 The overall scheme design shall be made for the observation station/network for reservoir-induced earthquake and the station/network shall be completed and put into use at the latest one year before impoundment. The observation shall be continued up to more than 6 years after impoundment, in which the water level reaches the design normal water level for at least 3 times. If a reservoir-induced earthquake occurs, observation work shall be continued. If no more reservoir-induced earthquake occurs, observation may be finished after it is reported to the local seismic authority for the record.

5.1.3 If necessary, deformation observation of the main active fault zone may be planned in reservoir/dam area to analyze the relationship between deformation of active faults and reservoir-induced earthquake, and provide information and data for studying the development trend of reservoir-induced earthquake.

5.2 Arrangement of Observation Networks

5.2.1 The observation range of reservoir-induced earthquake includes the area nearby the main structures and the reservoir section where earthquake is likely to occur. For extremely large or important projects, the observation scope may be expanded as appropriate and should cover the earthquake control fault nearby the reservoir.

5.2.2 The observation network shall consist of at least four seismic stations and one network management center, and digital telemetry network should be used. The distribution of seismic stations should be uniform and reasonable, and the station spacing should be between 10 km to 30 km. The stations and network shall be mainly arranged in the reservoir area immediately upstream of the dam and the reservoir section where an induced earthquake may occur. The observation stations for key areas shall be sensitive to the M_L 0.5 earthquake, and the observation stations for general areas shall be sensitive to the M_L 1.0 earthquake. If a reservoir-induced earthquake occurs, mobile seismic stations shall be added according to the earthquake situation.

5.2.3 The site of a seismic station shall be selected according to the requirements of Article 4.3.4 of this specification.

5.3 Technical Requirements of Observation System

5.3.1 The seismic station shall meet the following technical requirements:

1 The seismic signal detection in station observation equipment shall be conducted mainly by the seismometer, which may include the three-component short period seismmometer or the three-component broadband seismmometer. The accelerometer which reflects macro seismic effect of the site may be added as required.

2 The area of an observation station room should not be less than 12 m², the size of a seismometer pier should be 1.0 m × 0.8 m, which should be parallel with or vertical to the geographic meridian. The seismometer pier should be 0.3 m to 0.6 m above the ground floor, and the isolation tank shall be set up around the pier, lightning protection measures shall be taken for the observation room and equipment. The observation room shall be equipped with uninterruptible power supply or solar power supply system.

5.3.2 The network management center shall meet the following technical requirements:

1 The network management center should be located in the place where communication is unobstructed, transportation is convenient and power supply is guaranteed. Lightning protection measures should be taken for structure and equipment.

2 The network management center shall be equipped with computers, server and other equipment to constitute the LAN system, and shall have the functions of real-time seismic data receiving, processing, online observation data storage, real-time data flow exchange, data sharing and service, human-machine analysis and processing of earthquake events, as well as the system operation monitoring and

alarm.

5.3.3 Data transmission shall meet the following technical requirements:

1 The wire transmission, wireless transmission, or their combination may be used as the network data transmission modes according to site conditions.

2 The data relay station shall be set up when direct transmission between the observation station and the network center is unavailable, or when the transmission bit error rate cannot meet the requirements.

5.4 Observation and Operation

5.4.1 The equipment of the station and the management center, communications equipment, software, shall be calibrated, debugged and inspected after the completion of the observation network, and the network can be put into operation after it is accepted.

5.4.2 The network management center shall be on duty around the clock, responsible for the management, inspection and maintenance of the network management center and the seismic stations, and for the duty log keeping.

5.4.3 The observation methods and requirements for reservoir-induced earthquake shall conform to Article 4.3.5 of this specification.

5.4.4 The strong motion observation shall comply with the current sector standard DL/T 5416, *Specification of Strong Motion Safety Monitoring for Hydraulic Structures*.

5.5 Data Collation and Analysis

5.5.1 The network management center is responsible for processing, backup, analysis and collation of the observation data.

5.5.2 The collation and analysis of reservoir-induced earthquake observation data shall conform to the requirements of Article 4.3.6 of this specification. In addition, it is necessary to determine whether there is any reservoir-induced earthquake based on the seismic activity change observed and recorded by the network before and after reservoir impoundment, then analyze the characteristics and development trend of reservoir-induced earthquake. If a deformation observation for active faults is arranged in the reservoir/ dam area, the relationship between the deformation of active faults and the reservoir-induced earthquake shall be analyzed.

6 Groundwater Observation

6.1 General Requirements

6.1.1 The groundwater observation is intended to identify the characteristics and dynamic situation of groundwater in a hydropower project area, and provide basic data for analyzing hydrogeological conditions.

6.1.2 Groundwater observation shall be performed based on comprehensive collection and analysis of the information on meteorology, hydrology, topography, engineering geology and hydrogeology, and structure layout in the project area and preliminary analysis of groundwater status and its influence on project construction.

6.1.3 Groundwater simple observation or dynamic observation shall be planned based on the investigation extend and requirements of engineering geological analysis at different investigation stages.

6.1.4 The groundwater observation duration in early investigation stages shall not be less than one hydrological year. The observation duration at construction stage may be determined based on the situation exposed in excavation and the analysis need of groundwater.

6.2 Observation Contents and Range

6.2.1 The groundwater observation include groundwater level, flow, water temperature, turbidity, and water chemistry, and it shall also include water level, quality, volume and temperature of surface water in the project area.

6.2.2 The groundwater observation covers underground rivers, springs, wells, pits, rivers, gullies, boreholes, and exploratory adits which are related to the project and groundwater occurring after excavation of slope, foundation, underground caverns, etc.

6.2.3 Groundwater observation shall be conducted concurrently with the observation of rainfall in order to understand their relationship. When there is a local weather station, the weather and rainfall data from the station may be used, but when the weather station is far away, simple rainfall observation points shall be set up in the research area.

6.3 Observation Arrangement

6.3.1 The dynamic groundwater observation network consists of observation points and lines. The long-term observation boreholes shall generally be arranged along the flow direction of groundwater. The observation lines are generally arranged vertical or parallel to the flow direction of a river. The observation lines far away from the river should be arranged parallel or vertical to the groundwater flow direction or the micro-geomorphic boundary.

6.3.2 The observation lines should be arranged for dynamic groundwater observation in loose strata of a plain area. The spacing between observation points may be set from 200 m to 1000 m, and the number of observation points on the main observation line should not be less than 3.

6.3.3 Dynamic groundwater observation shall be arranged in the reservoir area including leakage areas, submerged areas, unstable slopes, and potential unstable slopes, dam area, and main structures area according to the following requirements:

1 At least 2 observation boreholes shall be arranged in the potential reservoir leakage area.

2 At least 2 observation boreholes shall be arranged in each geomorphic unit of a reservoir submerged area near and above the normal storage level.

3 The dynamic groundwater observation points in unstable or potential unstable slope should be arranged in conjunction with the slope deformation observation points, and the dynamic changes of surface water level should be observed simultaneously. Generally, no less than three investigation boreholes shall be selected as groundwater observation boreholes. The observation boreholes shall be arranged in the groundwater recharge area, runoff area, and discharge area.

4 The boreholes at both banks of the alternative dam sites and the selected dam site and in the main structures area should be taken as the dynamic groundwater observation boreholes. The observation points shall be arranged in different sections of an exploratory adit or shaft according to the hydraulic connectivity characteristics. In a dam area, the groundwater observation boreholes shall be arranged at different elevations. In alpine gorge area where groundwater is deeply buried, the groundwater level observation boreholes at both banks of the dam site may be arranged in exploratory adits.

5 For a pumped storage power station, the long-term groundwater level observation boreholes shall be arranged around the rim of the upper reservoir, especially at the saddles, thin ridges, and the slopes where the groundwater level is lower than the normal pool level. Long-term groundwater level observation boreholes should be arranged around the rim of the lower reservoir of a pumped storage power station or

the rim of the reservoir of a conventional hydropower station, when there are adjacent valleys, river bends, karst springs lower than normal pool level and geological structures hydraulically connecting the inside and outside of the reservoir. Long-term groundwater level observation boreholes shall be arranged along the waterway.

6   In karst areas, besides conventional groundwater observation, groundwater tracer testing shall also be carried out in order to analyze the relation of groundwater to rainfall and surface water, and study the characteristics of the seepage field, chemical field, temperature field and isotopic field of groundwater as required.

6.3.4 Observation points shall be set at natural springs, wells, pits, and boreholes in the project area and main seeping points in exploratory adits.

6.3.5 Observation points shall be set on the upper, middle and lower reaches of an underground river, natural river, or gully to understand the groundwater changes.

6.3.6 Convenience for access, management and observations shall be considered in selection of observation points. In dynamic groundwater observation area, hydrometeorological data shall be collected or water gauge be set to observe the groundwater level depending on the specific circumstances.

6.4 Observation Methods

6.4.1 Installation of measuring tools in observation boreholes shall meet the following requirements:

1   The observation boreholes in overburden should be drilled with casing, the observation boreholes in bedrock should be drilled with clear water, and the observation boreholes shall be carefully flushed before installation.

2   Segmented observation boreholes shall be water sealed strictly and the effects of packs shall be checked. The observation tubes may be set in parallel or concentric in the same borehole.

3   The observation borehole in a loose stratum shall be provided with a strainer tube, which is backfilled with sand and gravel in a thickness of not less than 20 cm. The strainer tube shall be aligned with the observed position of the aquifer.

4   Fixed markings and protective devices shall be provided at the collars of the observation boreholes.

6.4.2 Spring water observation instrument shall meet the following requirements:

  1. Water diversion and measuring devices should be set at the spring outlet, and the diversion device shall not affect the observation accuracy of water gushing, water temperature and water quality. Where it is difficult to build water diversion facilities, simple water measuring structures shall be set up, such as flow-measuring pools or weirs.

  2. A flow observation station like a canal or diversion canal may be set where a group of springs flow together in a straight section.

6.4.3 The observation methods of groundwater level shall meet the following requirements:

  1. Regular observation and recording may be conducted for simple observation of groundwater level by a measuring rope and bell, and an observation reference point for all observations shall be fixed at the same position of the well collar. Either a piezometer or an automatic water level gauge may be used for accurate observation to continuously record the water level change over time.

  2. The nearby surface water level shall be observed while the groundwater level is observed, and the precipitation data in the same period shall be collected.

  3. The initial water levels of pits, wells, and boreholes in loose strata shall be observed before water being fed into the borehole.

  4. In the drilling process, if water suddenly disappears or suddenly overflows, the drilling depth at that time shall be recorded timely.

  5. The water level of finished boreholes shall be observed after the residual drilling water being removed and before the borehole being sealed.

  6. The confined water head of artesian borehole should be observed by a connecting tube or a pressure gauge, and corresponding orifice device shall be set. Water gushing shall be measured at the same time, and water gushing test may be carried out if necessary.

  7. When two or more aquifers are found during drilling, the drilling operation shall be suspended and the borehole structure shall be redesigned in time, and packed for isolation. The stable water level of each aquifer shall be observed.

  8. Seal up the confined water at different boreholes depths when a confined aquifer is discovered during drilling. Identify the top

and bottom of confined aquifer, measure the water head and water temperature of confined water, then sample water for quality analysis.

6.4.4 Observation methods of groundwater quantity shall meet the following requirements:

1 For measuring low-flow groundwater, a cup or barrel and stopwatch may be used. For measuring large-flow groundwater in springs, gullies, adits, boreholes, and wells, a flow meter or weir should be used, and the flow rate measurement method may also be used.

2 When a measuring weir is used, different weir shapes shall be selected based on flow rate. A triangle weir is suitable for a flow rate of 1 L/s to 70 L/s, a rectangular weir is suitable for a flow rate greater than 50 L/s, and a trapezoidal weir is suitable for a flow rate of 10 L/s to 300 L/s.

3 For the site of seeping groundwater in an exploratory adit, changes of flow shall be measured individually or collectively depending on the hydraulic connectivity.

6.4.5 The groundwater quality observation shall meet the following requirements:

1 The groundwater quality observation is carried out by sampling.

2 The representative water samples are taken from each aquifer or seeping point for simplified water quality analysis. At least 3 groups of samples should be taken each time.

3 The water samples from long-term observation points should be taken quarterly, and at least one water sampling shall be done respectively in flood and dry seasons each year.

4 The water bottle shall be washed and cleaned before use, and rinsed with the sampled water before sampling.

5 Water shall be pumped to wash the borehole before sampling, the volume of pumped water shall be greater than two times the volume of water column in the borehole, and water samples shall be taken from the borehole after water level is restored.

6 When water samples are taken from exploratory wells, wells, springs, rivers, lakes or ponds, the sampling locations and positions shall be representative.

7 The water samples used for corrosive carbon dioxide analysis shall be filled into two bottles, and one of the bottles shall be added with

calcium carbonate powder.

8   In addition to the above observations, isotopic observation may be performed for groundwater in karst areas as required.

9   If the total analysis of water quality is necessary during engineering investigation, the sampling location, sampling quantity and test items and so on shall be determined based on the project requirements.

6.4.6   The observation methods of groundwater temperature shall meet the following requirements:

1   The thermometer that meets the observation accuracy requirements should be used for groundwater temperature observation.

2   Groundwater temperature should be observed in an aquifer, and the observation location for all observations should be same.

3   The atmospheric temperature shall be measured when water temperature is observed.

6.5   Observation Frequency and Accuracy

6.5.1   The groundwater observation equipment shall be calibrated regularly.

6.5.2   The elevation of groundwater observation points should not be inferior to the fifth-order leveling accuracy.

6.5.3   The observation frequency and accuracy of groundwater level shall meet the following requirements:

1   The observed groundwater levels are recorded in meters, and shall be accurate to centimeters (the second decimal place). Each observation shall be repeated twice, the difference between the values of two observations shall be less than 5 cm, and the average of two observation values shall be taken as the final observation value.

2   The long-term observation point of groundwater level should be observed once a week and, after a hydrological year, may be observed once a fortnight. The observation frequency shall be increased in rainy season or flood period, and the particularly important observation point shall be observed every one to two days after heavy rain, or instruments be used for automatic observation.

3   Groundwater level during drilling shall be observed once respectively after drill-lifting and before drill-lowering at shift change. Observation shall be conducted once every 12 h in the period when drilling is suspended for a long time.

4   The stable groundwater level in a finished borehole shall be observed after residual water in the borehole is removed. The groundwater level shall be observed once every 30 min, and be continuously observed 4 times or more. Until the water level variation in two adjacent observations in the last 4 consecutive water level observations is not greater than 5 cm, and no continuous change is found, the water level may be deemed stable, and the average value of the last two observations shall be taken as the stable water level of the finished borehole.

6.5.4   The observation frequency and accuracy of groundwater flow rate shall meet the following requirements:

1   The flow rate of groundwater seeping point in the project area shall be regularly observed. Generally, the observation frequency is once a month, but after rainfall, especially heavy rain, the observation frequency shall be increased.

2   The observation frequency for water bursting points exposed in an adit or a well shall be increased, and the dynamic groundwater observation at nearby observation points shall be increased accordingly.

3   The accuracy of groundwater flow rate observation should be controlled within ±5 % of the average of the observed flow values.

6.5.5   When the chemical composition of groundwater is unstable, sampling frequency shall be increased.

6.5.6   The observation frequency and accuracy of groundwater temperature shall meet the following requirements:

1   The number of water temperature observation points at the same aquifer shall not be less than 10 % of the groundwater level of the observation points.

2   The initial observation of groundwater temperature should be carried out simultaneously with the groundwater level observation, after certain law is known, the observation frequency may be reduced appropriately. For observation of geothermal water, the observation frequency may be determined based on the purpose and requirements of observation.

3   Groundwater temperature shall be measured after the thermometer reading is stable, and the reading shall be accurate to 0.5 °C.

4   Each groundwater temperature observation shall be repeated twice, and the difference between two observed values shall not be greater than

0.5 °C, then their average value shall be taken as the final value.

6.6 Data Collation and Analysis

6.6.1 The raw observation data and recording sheets of groundwater level, water quality, water temperature, flow, etc., shall be checked and verified at any time in order to find problems and solve them timely.

6.6.2 Routine processing of observation data shall include:

1. Calculate the physical indicators of dynamic elements including groundwater level, water quality, water temperature, flow, precipitation, surface water level, and water chemistry. based on the checked qualified observation data.

2. Plot the hydrographs, distribution graphs and relationship curves of dynamic elements.

3. Carry out statistics of characteristic values for dynamic elements such as maximum value, minimum value, variation, period, annual average value and annual variability.

6.6.3 The observation data shall be analyzed in time, the contents of analysis include the temporospatial change law of dynamic elements as well as the law and correlation of statistical characteristic values. The stability and trend values of dynamic elements change shall be analyzed and determined.

6.6.4 The charts and tables of results should include:

1. Observation results table of dynamic elements.

2. Statistical table of characteristic values of groundwater level and spring flow.

3. Hydrograph, distribution graph and various relation curves of dynamic groundwater elements.

4. Comprehensive graph of dynamic element change over years.

5. Distribution plan of observation network.

6. Hydrogeological section of observation line.

7. Hydrogeological section of the project area.

8. Groundwater level isograms as required.

6.6.5 Data collation instruction (or statement) should include:

1. General of data collation contents, methods and workload.

2. Adjustment and change of observation network.

3　General of observation methods, accuracy and measuring tool calibration.

4　Quality assessment of observation data.

5　Analysis results and conclusions of observation data.

6　Problems and suggestions.

7 Slope Deformation Observation

7.1 General Requirements

7.1.1 The slope deformation observation is performed to obtain the geological information and slope stability situation, check the stability analysis result and treatment effects, and predict the deformation development trend for project design, safe construction and operation.

7.1.2 The objects of slope deformation observation include unstable or potential unstable natural and cut slopes which would affect the project construction.

7.1.3 The slope deformation observation shall have clear purpose and prominent emphasis, which shall focus on the overall stability of slope, and also consider its local stability.

7.1.4 The slope deformation shall be observed by a combination of geological patrol inspection, simple observation and instrumental observation, and a combination of surface and subsurface observations.

7.1.5 Slope engineering geological observation shall be carried out based on the collection and analysis of slope deformation process and geological investigation data, taking into account the project needs. Special observation technical requirements shall be prepared if necessary.

7.1.6 When the slope deformation has occurred, observation points, lines or network shall be arranged for dynamic observation during geological investigation.

7.1.7 During slope deformation observation, surface water and groundwater shall be observed at the same time, and special observation shall be performed on precipitation, ground stress, and earthquake if necessary.

7.2 Observation Items and Contents

7.2.1 The items and contents of slope deformation observation mainly include slope displacement, length and opening of cracks appearing in slopes and structures, subsurface deformation, sliding surface, and fault activity.

7.2.2 Observation items and contents shall be selected according to the scale, importance and hazard of the slope. Corresponding observation points, lines or network shall be arranged.

7.3 Observation Arrangement

7.3.1 The slope deformation observation points, lines and network shall be arranged according to the following requirements:

1 For monitoring the displacement of key parts, observation points may be arranged, and the number of observation points should not be less than three.

2 For monitoring local deformation, one or more observation lines may be arranged along the direction of the main local deformation and the section parallel to it.

3 For monitoring the overall deformation of the slope and its affected area, an observation network composed of the observation lines intersecting longitudinally and transversely needs to be arranged to cover the whole area. The number of longitudinal observation lines should not be less than three, and that of transverse observation lines should not be less than two.

7.3.2 The observation points shall be arranged at the typical locations with engineering geological significance, such as the rear, middle and front of the slope, and the sensitive locations controlling sliding or deformation. The datum points of the observation lines and network must be set in the stable zone outside of the area influenced by slope deformation. The observation points shall be firmly attached to the rock and soil of slope, so that the observed value can represent the slope deformation.

7.3.3 The slope crack observation points may be arranged at the typical largest crack and on both sides of the potential fracture plane. The spacing of observation points should be 20 m to 30 m.

7.3.4 The observation lines shall be arranged in conjunction with the existing exploratory sections and stability analysis sections whenever practical, and at least one observation line shall be arranged in parallel to the main slope deformation direction. On each observation line, at least one observation point shall be located on the stable rock/soil mass. The distribution density and spacing of observation points shall be determined by the scale, distribution area and geometry of the slope. For a high slope with overall instability or potential instability or a large landslide, the spacing between two adjacent observation points may be 50 m to 400 m, and at least three points should be set on each observation line.

7.3.5 The surface observation network may be a cross grid, square grid, radial network, baseline intersection network, or random square grid, depending on the topographical and geological conditions.

7.3.6 The subsurface deformation observation points shall be arranged in combination with the surface deformation observation. Generally, the exploratory boreholes on the observation lines shall be used to arrange observation points, which shall be set in the positions with large deformation and possible failure. The number of observation boreholes on the main observation line should not be less than three, and the observation boreholes shall go through the potential sliding surface for more than 5 m. Observation points may be arranged for monitoring potential sliding surfaces or sliding zones at the exploratory adit revealing them.

7.4 Observation Methods

7.4.1 The slope deformation observation methods include geological patrol inspection, simple observation, geodetic survey, and embedded instrument observation.

7.4.2 Geological patrol inspection should include:

1. Occurrence and development of cracks in the ground surface and exploratory adits, and the displacement of the slope deformation boundary or discontinuities controlling the slope stability.

2. Changes in the number of springs at or nearby the slope; changes in flow rate, temperature and turbidity of spring water, gully water, and adit water; changes in groundwater level in boreholes and wells.

3. Deformation of project structures and residential houses on the slope.

4. Deformation of gully slopes, cut slopes, and embankment slopes.

5. Changes in frequency and amount of rock collapse at cliffs or high and steep slopes; phenomena such as bulging, creeping, piping, soil flowing in slopes.

6. Changes in the number and volume of landslides and collapses caused by rainstorm or flood.

7. The state of the rock mass liable to weathering and disintegration in cut slopes; deterioration due to water seepage in weak zones and special strata.

8. Cracking of shotcrete layers in cut slopes.

7.4.3 Geological patrol inspection shall meet the following requirements:

1. Geological patrol inspections include routine inspection and temporary inspection in case of danger, and shall be organized as required at investigation stage and construction stage.

2. Each geological patrol inspection shall be considered as a specific

geological investigation, performing fixed-point and follow-up observations and detailed field records.

3   The field records should be attached with geological sketches and photographs, and videos if necessary.

4   The field records shall be timely processed. Meanwhile, the overall or local stability of unstable slopes and landslides shall be analyzed in time.

5   Geological patrol inspection shall be summarized at least once a year, and reports and drawings shall be prepared accordingly.

7.4.4   For simple observation, simple marks and common measuring tools may be used in observing the cracks and subsurface displacement. The observation methods may be selected according to the following requirements:

1   For surface cracks, the steel tape or ruler measurement method, plumb observation method, and simple marking method may be adopted.

2   For subsurface displacement, the plastic tube-steel bar method, shaft concrete or steel ring method, and simple adit marking method may be adopted.

7.4.5   The geodetic survey method for surface displacement observation shall meet the following requirements:

1   When horizontal displacement is measured by the collimation line method, the width of the unstable slope should not be greater than 800 m, the visibility in the width direction shall be good, and there are alternative stable stations at both ends of collimation line. The observation method shall comply with the current sector standard DL/T 5178, *Technical Specification for Concrete Dam Safety Monitoring*.

2   Where topographical conditions are complex, the horizontal displacement should be measured by the triangulateration network or the intersection method. The observation method shall comply with the current sector standard DL/T 5178, *Technical Specification for Concrete Dam Safety Monitoring*.

3   Where conditions permit, the horizontal displacement of the ground surface may be observed by GPS.

4   Vertical displacement should be measured by the precise leveling method. The connecting levelling of the observation points and datum points shall conform to the national second-order leveling accuracy, and specific levelling shall comply with the current national standard

GB/T 12897, *Specifications for the First and Second Order Leveling*.

5   Vertical displacement may also be measured by the trigonometric elevation method. The specific measurement shall comply with the current sector standard DL/T 5178, *Technical Specification for Concrete Dam Safety Monitoring*.

7.4.6   The embedded instrument observation method shall meet the following requirements:

1   The instruments for slope deformation observation shall be less but better. The selected instruments and equipment shall meet the requirements of observation accuracy, should have long-term stability and reliability, as well as easy use and maintenance. The check and calibration of instruments and equipment shall comply with the current sector standard DL/T 5178, *Technical Specification for Concrete Dam Safety Monitoring*.

2   Subsurface displacement may be observed by borehole inclinometer, movable inclinometer, borehole multi-point displacement meter and extensometer (convergence meter), and the time domain reflectometry (TDR) may also be used, to detect and measure the location and interface displacement of deep-seated sliding surface. The borehole inclinometer casing may be made of aluminum alloy. The borehole shall reach to about 5 m below the main sliding surface. The main measurement direction shall be basically consistent with the expected main sliding direction. The observation line direction of the multi-point displacement meter and extensometer shall also be consistent with the main deep-seated sliding direction.

3   The multi-point displacement meter and the borehole inclinometer should not be used together at the same location. The multi-point displacement meter should be arranged in the slope with outcropping faults, interlayers or bedding planes, and the borehole shall pass through the weak discontinuities to be monitored. The displacement observation points inside the soil mass where the borehole inclinometer is used may be set within the predicted sliding range in combination with the longitudinal observation lines for surface displacement. The deep-seated displacement observation points on one longitudinal observation section should not be less than two.

4   In exploratory adits revealing unstable slopes or sliding surfaces (sliding zones) of landslides, the multi-point displacement meter and convergence meter may be used to observe the sliding displacement

along the sliding surface or sliding zone.

5 For rock slopes, acoustic emission monitoring may be arranged if required, to analyze the rock mass failure degree based on the frequency and amplitude.

7.4.7 The use of positive and negative signs for deformation observation data shall meet the following requirements:

1 Horizontal displacements: positive for main sliding direction, negative for opposite.

2 Vertical displacement: positive for drop, negative for rise.

3 Cracking: positive for opening, negative for closing.

7.5 Observation Frequency and Accuracy

7.5.1 The frequency of geological patrol inspection and simple observation shall be determined according to the slope stability. For the slopes with good stability and without obvious deformation, the observation frequency may be once a month normally, and once every two weeks in rainy season or flood period, and the observation should be conducted in time after rainstorm or strong earthquake. For the slopes under deformation or with obvious deformation, the observation frequency shall be once a week, and shall be increased in case of abnormality.

7.5.2 The observation frequency of geodetic survey shall be determined according to the slope stability. For the slopes with good stability and no obvious deformation, the observation frequency may be once every two months normally, and once every month in rainy season or flood period, and the observation should be conducted in time after rainstorm or strong earthquake. For the slopes with obvious deformation, the frequency shall be once a week, and shall be increased if obvious abnormality of horizontal or vertical displacement is found. The stability of the datum point should be checked every half a year, and that of the datum network should be checked once a year.

7.5.3 The deformation observation accuracy shall meet the following requirements:

1 For horizontal displacement at observation points, the allowable observation error is ±3 mm, or 1/5 to 1/10 of the annual displacement.

2 For vertical displacement at observation points, the allowable observation error is ±3 mm.

3 For crack opening measurement, the allowable error is ±0.5 mm.

7.6 Data Collation and Analysis

7.6.1 Observation data collation and analysis shall include the following:

1. Check the raw data.

2. Routinely process and analyze the observation data, and the abnormity report.

3. Regularly prepare the observation analysis report.

7.6.2 The raw observation data shall be checked as follows and the problems found shall be timely solved:

1. Whether the site operation methods meet requirements.

2. Whether the checked observation data are within the allowable error.

3. Whether the recorded data are accurate, complete and clear.

7.6.3 The routine processing of observation data shall include:

1. Calculate the observed physical indicators such as horizontal displacement, vertical displacement, crack opening based on the qualified observation data, and record them in the corresponding forms.

2. Plot the process curves, distribution maps and the deformation curves for the observed physical indicators.

3. Perform the statistics on the characteristic values of observed physical indicators, including maximum and minimum values (with occurrence time), amplitude, average, rate of change, etc.

7.6.4 The observed physical indicators shall be analyzed in time to obtain:

1. Variation pattern of the observed physical indicators with time and space.

2. Regularity of the statistical characteristic values of the observed physical indicators.

3. Correlation between the observed physical indicators.

4. Stability and trend of variation of the observed physical indicators, and the potential adverse effects on project in the future.

7.6.5 When any abnormality is found in the observation data, and confirmed potential hazard, early warning and suggestions on engineering treatment measures shall be given in time.

7.6.6 The periodic observation report shall include:

1. The project planning and design, current stage of geological

investigation and design, and accomplished geological investigation workload.

2 Overview of topography and geology.

3 Overview of slope deformation observation, including arrangement, maintenance, serviceability rate and modification of the observation network and points, as well as the calibration of instruments, equipment and measuring tools.

4 Charts of results compiled from observation data.

5 Observation situations and main results of geological patrol inspection.

6 Description of observation data processing methods.

7 Analysis results and conclusions of observation data.

8 Problems and suggestions.

8 Unstable Rock Mass Observation

8.1 General Requirements

8.1.1 The purpose of unstable rock mass observation is to obtain the observation data and analyze the failure modes (toppling, slippage, rotation, subsidence, opening, etc.), and the deformation direction and rate, predict the stability, and put forward treatment measures and suggestions.

8.1.2 For the unstable rock mass which endangers the safety of personnel and project construction, corresponding observation shall be arranged according to the rock mass scale and the degree of influence on the project. Refer to Appendix C of this specification for the scales and main types of unstable rock mass.

8.1.3 The topographical and geological data about unstable rock mass and the surrounding areas shall be collected to analyze the boundary and stability condition and carry out observation of the unstable rock mass.

8.1.4 The factors, such as rainfall, surface water, groundwater, earthquakes and human activities that would influence the stability of unstable rock mass and their related data shall be collected to analyze and evaluate their impacts on the stability of unstable rock mass. In the absence of related data, specific observation shall be arranged as required.

8.1.5 For the unstable rock mass with high hazards, observation points, lines or networks shall be arranged for dynamic observation during the geological investigation.

8.1.6 Both surface water and groundwater shall be observed simultaneously during deformation observation for the unstable rock mass; and the precipitation, ground stress, and seismicity shall be observed when necessary.

8.2 Observation Items and Contents

8.2.1 The engineering geological observation items and contents for unstable rock mass include the displacement, the length and opening of cracks, and their changes.

8.2.2 The deformation observation items and contents shall be selected by the scale, importance and hazard degree of unstable rock mass, and corresponding observation shall be arranged.

8.3 Observation Arrangement

8.3.1 For the unstable rock mass area that is accessible for both personnel and equipment, observation points, lines or network shall be arranged based on the type and scale of unstable rock mass:

1 For the unstable rock mass of large or medium scale, observation points may be arranged at the key points, which should not be less than two. For the unstable rock mass of extra-large scale, observation points, lines or network may be arranged as per actual situation or shall be arranged according to Article 7.3.1 of this specification.

2 The arrangement of observation points shall be compatible with the type of unstable rock mass. For the unstable rock mass with toppling, tensile cracking and buckling, observation points should be arranged at the rear of cutting surface and unstable rock mass. For the unstable rock mass with slipping and tension cracking, observation points should be arranged at weak rock mass where deformation occurs in lower parts. For the isolating unstable rock mass, observation points should be arranged at unstable rock mass.

3 The observation points for cracks may be arranged at the representative largest cracks and on both sides of the potential fracture planes. The spacing of observation points should be within 10 m.

8.3.2 For the unstable rock mass on the cliff that is too high and steep to climb, observation datum should be arranged at the opposite bank of the unstable rock mass or a visible and appropriate location.

8.4 Observation Methods

8.4.1 For the unstable rock mass area that is accessible for personnel and equipment, the observation method shall be selected according to the scale and hazard degree of the unstable rock mass and on the following principles:

1 For unstable rock mass of large scale, extra-large scale or with significant hazard, observation methods shall be selected based on the environmental conditions and the technical feasibility, including geological patrol inspection, simple observation, geodetic survey and instrumental observation. The specific method shall be in accordance with Articles 7.4.2 to 7.4.7 of this specification.

2 For medium or small scale unstable rock mass, the main observation methods shall be geological patrol inspection, and simple observation.

8.4.2 For the unstable rock mass on the cliff that is too high and steep to climb, rock mass deformation may be observed by close-range photogrammetry, such as three-dimensional laser scanning or digital photography.

8.4.3 The requirements for observation frequency and accuracy of unstable rock mass shall be in accordance with Articles 7.5.1 to 7.5.3 of this specification.

8.5 Data Collation and Analysis

8.5.1 The original observation records shall be checked mainly to confirm the correctness of the observation methods and the completeness of the observation data. The site check shall be performed if there is a significant variation between the current and previous observation data.

8.5.2 Draw observation diagrams based on observation data, and analyze the change pattern of observation data with time periodically, as well as the correlation and internal connection of observation data.

8.5.3 In case of an abnormality in the observation data, and deformation aggravation is confirmed, early warning and suggestions on engineering treatment measures shall be given in time.

8.5.4 Observation and analysis reports shall be prepared periodically, which should include:

1　Project overview.

2　Description of the design stage, and the accomplished geological investigation workload.

3　Topographical and geological conditions in the work area.

4　Distribution and basic characteristics of unstable rock mass, and analysis of the factors affecting stability.

5　Observation arrangement and methods.

6　Observation results and stability evaluation of unstable rock mass, and analysis of its influence on the project.

7　Suggestions on treatment and observation of unstable rock mass.

9 Debris Flow Observation

9.1 General Requirements

9.1.1 The purpose of engineering geological observation of debris flow is to obtain observation data, analyze and predict the development trend and hazard degree, and provide the basis for prevention and control of debris flow.

9.1.2 For the debris flow which endangers the safety of the personnel and project construction, corresponding observation shall be arranged according to the scale and type of debris flow and the degree of influence on the project. The classification of debris flow may be in accordance with Appendix D of this specification.

9.1.3 The engineering geological observation of debris flow shall be carried out based on collection of meteorologic, hydrologic, topographic and geologic data in the river basin where debris flow occurs, and analysis of forming conditions, movement, fluid characteristics, and accumulation features of debris flow.

9.1.4 For the debris flow with high hazard, observation points shall be arranged for dynamic observation.

9.2 Observation Items and Contents

9.2.1 The engineering geological observation items and contents of debris flow shall comply with Table 9.2.1:

Table 9.2.1 Engineering geological observation items and contents of debris flow

No.	Observation item	Observation content
1	Source of solid material	Stability of slumped mass; Stability of loose deposits subjected to scouring by rainstorm and flood
2	Precipitation	Rainfall and its duration; Water volume and duration of melting when water source is ice, snow and frozen soil; Seepage when there is a mountain lake, reservoir or channel upstream or in high land; Rainfall infiltration and groundwater dynamics in the section with concentrated distribution of solid materials
3	Dynamic observation	Debris level, flow velocity, flow regime and rise and drop process
4	Flow in gully	Flow of surface water in different locations of the gully

9.2.2 The observation items and contents shall be selected based on the scale and hazard degree of debris flow, and corresponding observation facilities shall be arranged.

9.3 Observation Arrangement

9.3.1 The observation arrangement for solid material source of debris flow shall meet the following requirements:

1. If the solid material sources from slumped mass, the deformation observation points, lines and network shall be arranged according to Section 7.3 of this specification.

2. If the solid material sources from loose deposits, the stability observation points and network shall be arranged based on the zoning by erosion degree, and the density of stability observation points for loose deposits may be in accordance with Table 9.3.1, and the identification of erosion degree shall comply with Appendix D of this specification.

Table 9.3.1 Density control of observation points for stability of loose deposits

Erosion degree	Observation point density (point/km^2)
Extremely severe eroded area, Severe eroded area	20 - 30
Moderately eroded area	10 - 20
Slightly eroded area Not obviously eroded area	0 - 10

NOTE The number of observation points may be adjusted based on the developing trend and change of erosion degree.

9.3.2 The precipitation observation points shall be arranged according to the following requirements:

1. The observation points shall be arranged in the representative area in a debris flow gully or in a river basin, mainly in the forming area of debris flow.

2. The observation points shall be arranged in the area where it is relatively flat and less affected by wind. In general, the distance between the instrument and the obstacle shall not be less than twice the height difference between the obstacle vertex and the instrument.

3   The number of observation points depends on the area of the river basin where debris flow gully is located and the representativeness of observation points, and shall not be less than three generally. The observation points may be arranged in network for the debris flow gully with a large catchment area, otherwise, the points may be arranged in triangle.

9.3.3 Dynamic observation of debris flow shall be arranged according to the following requirements:

1   The dynamic observation points shall be generally arranged in the area with a small scouring and deposition change, and mainly in flowing area.

2   The dynamic observation points shall be set in stable positions on both banks.

3   The number and spacing of the dynamic observation sections should be determined according to the topographical and geological conditions of gully, and generally, the number of observation sections should not be less than two.

9.3.4 The gully flow observation should be arranged in the collecting area, flowing area and outfall of gully water, and generally, the number of observation sections should not be less than two. The observation section shall be added where the upstream and downstream gully water changes obviously.

9.4 Observation Methods

9.4.1 The observation methods for solid material source of debris flow shall meet the following requirements:

1   When the solid material of debris flow sources from slumped mass, the observation methods shall comply with Section 7.4 of this specification.

2   When the solid material of debris flow sources from loose deposits, standard sheet erosion observation fields may be set in the areas with different geological conditions to observe the erosion amount in different rainfall conditions, and analyze the solid materials amount under the critical rainfall in which debris flow is formed.

9.4.2 The precipitation observation methods shall meet the following requirements:

1   For rainstorm-induced debris flow, weather station shall be set up mainly to observe rainfall.

2   For meltwater-induced debris flow, the monitoring for melting amount of ice and snow shall be added.

9.4.3   Water flow in gully may be observed by staff gauge, automatic recording device of water level and flow, measuring weir, measuring rod, etc. to observe the dynamic changes in level and flow of gully water.

9.4.4   The observation frequency is generally once a month, and once every 15 days in flood season. In case of danger, observation should be carried out once a day, and continuous observation shall be done if necessary.

9.5   Data Collation and Analysis

9.5.1   For original observation records, the correctness of observation methods and completeness of observation data shall be checked. Analysis shall be carried out if there is a significant variation between the current and previous observation data.

9.5.2   Draw observation diagrams based on observation data and analyze the change pattern of observation data with time periodically, as well as the correlation and internal connection between observation data.

9.5.3   In case of abnormality in the observation data of debris flow and potential hazard is confirmed, early warning and suggestions on treatment measures shall be given timely.

9.5.4   Observation and analysis reports shall be prepared periodically, which should include:

1   Project overview, description of design stage, and the accomplished geological investigation workload.

2   Natural environment conditions and human activity situation in working area.

3   Analysis on basic geological characteristics of the river basin, and the forming conditions of debris flow.

4   Observation arrangement and methods.

5   Analysis of observation data.

6   Analysis of basic characteristics of debris flow.

7   Analysis of hazard degree and development trend of debris flow.

8   Conclusions and suggestions.

10 Geotechnical Observation on Foundations

10.1 General Requirements

10.1.1 The observation of soil and rock masses in foundation is intended to obtain geological information about foundations or foundation pits, to analyze and judge the safety of the foundations and foundation pits, and provide the basis for project design and construction.

10.1.2 The observation of soil and rock masses covers all types of foundations and foundation pits in hydropower projects, and the observation contents include abnormality and change trend of soil and rock masses after excavation. Based on the acquired information, suggestions on optimization and treatment are made.

10.1.3 Observation shall be performed based on collection of the information about topographical and geological conditions, design and construction of foundation and foundation pit.

10.2 Observation Items and Methods

10.2.1 For rock foundations, the observation contents shall include:

1 Deformation of rock mass in foundation including rebound and relaxation, heave, creep, cracking, and cracks in ground and structures around foundations.

2 Weathering rate of the rock mass liable to weathering, slaking rates of the rock mass liable to slaking, and deterioration of weak layers and water-soluble salt stratum due to seepage.

3 Deformation of slopes in foundation pits.

4 Outcrop location, type, flow, temperature, and color of groundwater, material outflowing with groundwater and deposition; water gushing, water level and their dynamic change in foundation pit.

5 Change of groundwater in foundations and foundation pits when the cofferdam or the dam is temporarily used for water retaining during the construction period.

10.2.2 For soil foundations, the observation contents shall include:

1 Deformation of foundation soil including heave, settlement, creep, extrusion, cracking, etc.

2 Cracking and seepage of water-stop curtain in foundations.

3 Cracking, settlement and slippage of soil behind the diaphragm wall.

4   Soil flow, sand gushing, piping in foundation pits, and slope deformation.

5   Outcrop location, type, flow, temperature, and color of groundwater, material outflowing with groundwater and deposit in foundation pits; water gushing, water level and their dynamic change in foundation pits; abnormality with drainage of surface water and groundwater at the site and dewatering in foundation pits.

6   Stockpiling around foundations, cracking and settlement of ground and structures around foundation.

10.2.3 Observation of rock and soil in foundations should adopt geological patrol inspection and simple observation methods. The observation methods shall comply with Articles 7.4.3 and 7.4.4 of this specification.

10.2.4 Observations of the foundation pit parameters such as the wall displacement, horizontal displacement of surrounding soil mass, vertical displacement of a column, lateral soil pressure of a retaining wall, pore water pressure and settlement of surrounding structures shall be performed by professional institutions.

10.3 Observation Frequency and Data Processing

10.3.1 During excavation of a foundation or a foundation pit, the observation shall be conducted once a day generally. When there is a considerable change with the observation results, the observation frequency shall be increased, and if there is a sign of accident, continuous observation shall be done.

10.3.2 The time intervals for different observation items may be determined according to the construction progress. The observation frequency at the non-excavation stage may be adjusted in case of abnormality.

10.3.3 Observation data shall be processed and analyzed routinely, and when the observation data varies greatly, it shall be reported timely. And if there is a sign of accident, early warning shall be given in time.

10.3.4 Observation analysis reports shall be prepared regularly, which should include:

1   Project overview, description of design stage, and the accomplished geological investigation workload.

2   Topographical and geological conditions of the observation areas.

3   Contents, layout and methods of observations of rock and soil mass of foundations.

4   Processing and analysis results of observation data.

5   Observation conclusions and foundation treatment suggestions.

6   Problems and improvement suggestions.

11 Surrounding Rock Observation in Underground Cavern

11.1 General Requirements

11.1.1 The observation of surrounding rocks in underground caverns is intended to obtain geological information about surrounding rocks in underground caverns to analyze the stability and safety of surrounding rocks, and provide the geological basis for design and construction.

11.1.2 Engineering geological observation shall be performed in one of the following:

1 Cavern and tunnel section with fault zone, closely jointed zone, weak zone, alteration zone, soft rock, swelling rock, saline rock, karst deposits, soil layer, or water soluble salt stratum.

2 Cavern and tunnel section with groundwater.

3 Cavern and tunnel section with ground temperature, harmful gas, or radioactive anomalies.

4 Underground caverns located in high ground stress zones.

5 Rock-bolted crane girders, rock pillars, rock walls in underground caverns.

6 Shallow-buried tunnels, gully-crossing (river-crossing) tunnel sections, and the tunnel sections with structures on the ground surface.

11.1.3 Observation of underground caverns and tunnels shall be performed based on collection and analysis of topographic and geological, design and construction data.

11.2 Observation Contents and Methods

11.2.1 The observation of surrounding rocks in underground caverns and tunnels shall include:

1 Deformation, collapse height and seepage-induced deterioration of surrounding rocks in the sections with fault zone, closely jointed zone, weak zone, alteration zone, soft rock, swelling rock, saline rock, karst deposits, water-soluble salt stratum, and soil layer.

2 Extension and slipping of discontinuities.

3 Flow, water temperature, material with water flowing, and water pressure at groundwater outcrop point.

4   Relaxation and fracture in the rock-bolted crane girders, rock pillars, and rock walls in underground caverns.

5   Relaxation, inner bulging, buckling, and squeezing of surrounding rock, and width of crack.

6   Failure mode, sound characteristics, movement feature, influence depth, block size and fracture failure characteristics, development and time duration of surrounding rocks under high ground stress.

7   Unstable blocks in underground caverns and tunnels.

8   Cracking and abnormalities in shotcreted and supported areas.

11.2.2 Observation of surrounding rocks in underground caverns should adopt the geological patrol inspection method combined with simple observation, and record and describe corresponding results in accordance with Article 11.2.1 of this specification. 3D laser scanning and digital photography may also be adopted.

11.2.3 The observation of the convergence and deformation, displacement, relaxed zone, water pressure and ground temperature of surrounding rocks, harmful gas and radioactivity testing shall be performed by professional institutions. The observation items for surrounding rocks in underground caverns may be in accordance with Appendix E of this specification.

11.3 Observation Frequency and Data Collation

11.3.1 During excavation of underground caverns, the observation is conducted generally once a day or once every two days. When the observation result varies greatly, the observation frequency shall be increased, and if there is a sign of accident, continuous observation shall be done.

11.3.2 The time intervals for different observation items may be determined based on the construction progress. The observation frequency of non-excavation stage may be adjusted based on abnormal situation.

11.3.3 Observation data shall be routinely processed and analyzed, and when the observation result varies greatly, it shall be reported timely. If there is a sign of accident, early warning shall be given in time.

11.3.4 The observation analysis reports shall be prepared regularly, which should include:

1   Project overview, and description of the investigation and design stage, and the accomplished geological investigation workload.

2   Overview of topographical and geological conditions.

3 Contents, layout and methods of observation of surrounding rocks.

4 Processing and analysis results of observation data.

5 Observation conclusions and surrounding rock treatment suggestions.

6 Problems and improvement suggestions.

Appendix A Engineering Geological Observation Items of Each Investigation Stage

Table A Engineering geological observation items of Each investigation stages

S.N.	Observation item	Planning	Prefeasibility study	Feasibility study	Bidding design	Detailed design
1	Fault and seismic activity	–	+	+	+	+
2	Reservoir-induced earthquake	–	–	–	+	+
3	Groundwater	+	√	√	√	√
4	Slope deformation	+	+	√	√	√
5	Unstable rock mass deformation	–	+	+	+	√
6	Debris flow	–	+	+	+	√
7	Geotechnical observation on foundation	–	–	–	–	√
8	Observation of surrounding rocks in underground cavern	–	–	+	+	√

NOTE "√" denotes the item that shall be observed; "+" denotes the item to be observed depending on actual need; "–" denotes the unnecessary observation item.

Appendix B Comprehensive Classification of Active Faults

Table B Comprehensive classification of active faults

Class		Comprehensive features (activity)	Annual average activity velocity V (mm/a)	Magnitude of historical earthquake and paleo-earthquake M	Trend
I	Strongly active faults	The faults have strong activity trend, and are lithosphere-order deep and major faults mostly, with a length of more than 100 km largely and 20 km at least. The faults have been active intensively and persistently since Quaternary Period, which were active in Late Pleistocene Epoch, and became strong in Holocene Epoch. Therefore, the faults manifest as seismogenic structure of strong earthquakes from ancient times to present	$V \geq 10$	$M \geq 7$	Strong earthquake likely to occur
II	Moderately active faults	Different activity trends are found in different fault sections, some creep slowly, while others have strong activity, fault displacement occurs at moderate velocity. These faults are mostly of active ones in the base with a length of 20 km to 100 km. The faults have been active persistently since Quaternary Period, which were active in Late Pleistocene Epoch, and became strong in Holocene Epoch. Therefore, the faults manifest as seismogenic structures of moderate to strong earthquakes from ancient times to present	$1.0 \leq V < 10$	$5 \leq M < 7$	Moderate earthquake likely to occur

Table B *(continued)*

Class		Comprehensive features (activity)	Annual average activity velocity V (mm/a)	Magnitude of historical earthquake and paleo-earthquake M	Trend
Ⅲ	Weakly active faults	The faults creep slowly, and their displacement is at low velocity. They are overburden faults with a length of tens of kilometers. The activity of faults has been weak and continued to decline since Quaternary Period, and was not obvious in Holocene Epoch. Therefore, the faults manifest as weak earthquake concentrated distribution zones from ancient times to present	$0.1 \leq V < 1.0$	$M < 5$	Weak earthquake likely to occur

Appendix C   Scales and Main Types of Unstable Rock Mass

C.0.1   The scale of unstable rock masses may be classified in accordance with Table C.0.1.

Table C.0.1   Scale classification of unstable rock masses

Scale	Small	Medium	Large	Extra large
Volume V (m³)	$V < 100$	$100 \leq V < 1000$	$1000 \leq V < 10000$	$V \geq 10000$

C.0.2   Types of unstable rock mass may be classified in accordance with Table C.0.2.

Table C.0.2   Type classification of unstable rock mass

Type		Discontinuity	Failure mechanism	Major instability factor
Toppling		Vertical joint, Columnar joint, upright or overturning stratum	Toppling	Water pressure, earthquake force, gravity
Sliding		Discontinuities toward free face	Sliding	Gravity, water pressure, earthquake force
Tensile cracking		Weathered fissure and tensile fracture due to gravity	Tensile	Gravity, earthquake force
Buckling		Steeply dipping and well developed fissure, discontinuities not toward free face	Buckling	Gravity, earthquake force
Combined	Toppling & sliding	Vertical joint, columnar joint, upright or overturned stratum, discontinuities toward free face	Toppling, sliding	Water pressure, earthquake force, gravity
	Toppling & tensile cracking	Vertical joint, columnar joint, upright or overturned stratum, weathered fissure and tensile fracture owing to gravity	Toppling, tensile	Water pressure, earthquake force, gravity

Table C.0.2 *(continued)*

Type		Discontinuity	Failure mechanism	Major instability factor
Combined	Fracturing & sliding	steep dip fissure well developed, discontinuities toward free face	Fracture, sliding	Gravity, water pressure, earthquake force
Isolating		Isolated rock blocks (group)	Rockfall	Gravity, earthquake force

Appendix D Classification of Debris Flow and Erosion Degree

D.0.1 Classification of debris flow may be determined in accordance with Table D.0.1.

Table D.0.1 Classification of debris flow

Basis	Indicator	Type of debris flow
Geomorphic features	Drainage area features	Gully Slope
Genetic types	Water source	Rainstorm-induced type Snow and ice melting type Dyke/dam failure type
	Material source	Rock fall and landslide Slope erosion Moraine deposits Pyroclastic Engineering disposal material Multi-sourced
Mechanical mechanism and fluid features	Formation Mechanism	Soil mechanical Hydraulic
	Fluid property	Diluted Viscous
Development	Frequency of occurrence	High frequency Moderate frequency Low frequency
	Scale	Extremely large Large Medium Small
	Forming time	Ancient Old Young
	Hazard	Extremely large Large Medium Small
	Potential risks	Extremely large Large Medium Small

D.0.2 Classification of debris flow by geomorphic features may be determined in accordance with Table D.0.2.

Table D.0.2 Classification of debris flow by geomorphic features

Basis	Type	
	Gully	Slope
Boundary and shape	With drainage basin as boundary, restrained by the valleys; the formation, transport, deposit of debris flow are relatively distinct; the outline is in dumbbell shape	There is no constant boundary or obvious gully but only changing boundary; the outline is in bowling shape
Failure mechanism	With gully as the center; loose deposits in material source area are distributed at both banks of the gully and gully bed; failure mechanism is mainly erosion of rainstorm and confluent water on material source	Debris flow occurs on the slope of over 30°, underlain by bedrock and weakly permeable layer; the material sources are mostly overburden; the scale is relatively small; failure mechanism approximates to collapse and slide
Identifiability and scale	The occurrence time and location are regular and identifiable, and the hazard scale is relatively large	It is not easy to identify occurrence time and location, and the hazard scale is relatively small
Recurrence	With large volume and long return period, it might occur successively and repeatedly	With small volume and long return period, it has less successiveness and repeatability
Preventability	Mostly distributed in belt or large area in the same region, it is identifiable and preventable	It is randomly distributed at several places on the same slope, arrayed like comb; upper side of the flow has a certain distance from the ridge. It is of low identifiability and preventability

D.0.3 Classification of debris flow by water source may be determined in accordance with Table D.0.3.

Table D.0.3 Classification of debris flow by water source

Type	Basis
Rainstorm-induced type	Formed under the action of rainfall and rainstorm in a long period, or under severe rainstorm
Snow and ice melting type	Induced by the erosion of the melt water of snow and ice on bank slope and valley bed, and sometimes together with the action of rainfall
Dyke/dam break type	Induced by sudden, high-intensity flood scouring, which is caused by the failures of barrages or dams due to water flow erosion, earthquake, dyke instability

D.0.4 Classification of debris flow by material source may be determined in accordance with Table D.0.4.

D.0.4 Classification of debris flow by material source

Type	Basis
Rock fall and landslide	Debris flow is directly formed by landslide or rock fall, or formed by the deposits from landslide or rock fall under the action of rainfall and other factors
Slope erosion	The solid materials come from slope and gully erosion deposit in the gully, and induce debris flow under the action of water flow
Moraine deposits	The main source of solid materials of debris flow is moraine deposits
Pyroclastic	The main source of solid materials of debris flow is pyroclastic accumulation
Engineering disposal material	The main source of solid materials of debris flow is waste soils and muck from engineering works, which is a typical man-made flow
Multi-sourced	A combination of two or more of the ones above, caused by joint action of nature and human activities

D.0.5 Classification of debris flow by formation mechanism may be determined in accordance with Table D.0.5.

Table D.0.5 Classification of debris flow by formation mechanism

Type	Soil mechanical	Hydraulic
Basis	Caused by soil slide, dislocation, collapse and falling	Mostly developed in valleys and year-round streams in mountainous areas, and activated by major flood

D.0.6 Classification of debris flow by fluid property may be determined in accordance with Table D.0.6.

Table D.0.6 Classification of debris flow by fluid property

Basis	Type	
	Diluted debris flow	Viscous debris flow
Composition of slurry content and features	The slurry contains no or less viscous material; with a viscosity of less than 3 Pa · s, without yield stresses; it is a Newtonian fluid	The slurry is rich in viscous material (clay and silt with particle size smaller than 0.01 mm); with a viscosity of greater than 3 Pa · s, reticular structure, with yield stresses; it is a non-Newtonian fluid
Composition of solid content and features	The coarse particles in solid content consist of gravel, blocks, coarse sand and a little silty sand	The coarse particles consist of gravel, blocks, and silty sand greater than 0.01 mm
Flow state	Intense turbulent flow, solid and liquid compositions move in different velocities, and exchange in vertical direction, with phenomena of plume and diffuse flows; debris flow has a large change in erosion and deposition; less mud residue	Pseudo single-phase laminar flow moves entirely sometimes; no exchange in vertical direction; slurry is sticky and high in uplift force; fluid has obvious drag reduction, with high capacity of forward movement and visible climbing up at corners; debris flow's erosion and sediment area alters invisibly along the traces; mud residue due to its good adhesion
Accumulation features	Deposits have certain sorting feature; there are fan-type deposits and the lateral dyke type stripped deposits in horizontal plane; a typical phenomenon of erosion on concave bank and sediment on convex bank can be seen in the bend; sediments are dominated by coarse grains	Deposits have no sorting feature, and the clay and gravel are mixed together; deposits sediment is like tongue in horizontal plane, remaining the structure characters when flowing; there are no visible bedding planes in the sediments, however, there are obvious sediment planes from different debris flow occurrence, with less chalazoidite
Bulk density	1.30 g/cm^3 (inclusive) to 1.80 g/cm^3	1.80 g/cm^3 (inclusive) to 2.30 g/cm^3

D.0.7 Classification of debris flow by forming time may be determined in accordance with Table D.0.7.

Table D.0.7 Classification of debris flow by forming time

Type	Young debris flow	Old debris flow	Ancient debris flow
Basis	Active debris flow in recent 100 years, which is judged by site investigation and observation on valley deposits, and the deposits are unconsolidated mostly	Inactive Debris flow formed 100 years ago. The debris flow deposits are not consolidated or underconsolidated	Debris flow did not occur in recent hundred years, but occurred in Quaternary Period. The debris flow deposits are fully consolidated or underconsolidated

D.0.8 Classification of debris flow by frequency of occurrence may be in accordance with Table D.0.8.

Table D.0.8 Classification of debris flow by frequency of occurrence

Type		High frequency	Moderate frequency	Low frequency
Basis	Debris flow action	The debris flow is different in scale, the valley bed is raised obviously, and older sediment materials can be seen often; in the formation area, collapse and landslides are developed, vegetation coverage is low, and erosion modulus is greater than 2000 t /(km²·a)	The debris flow is different in scale, the valley bed is raised obviously, and older sediment materials can be seen often; in the formation area, collapse and landslides are relatively developed, vegetation coverage is moderate, and erosion modulus is smaller than 2000 t / (km²·a)	The debris flow is relatively large in scale, the valley bed is cut because of intense erosion; older sediment fan remains in good condition, and the old terraces remains in good condition also, with large area of distribution. In the drainage area, landslides are inactive, vegetation coverage is high, and erosion modulus is often smaller than 2000 t / (km²·a)
	Debris flow activity and accumulation features	Debris flow occurs frequently, and delivers large amounts of sediment into the main river channel; the land in dangerous area is difficult for use	Debris flow occurs at intervals, and delivers much sediment into the main river channel; the land use in dangerous area is low	The flood marks of recent debris flow are not obvious, however, activity is intense and large scale, often causes significant disaster, therefore, this kind of debris flow is of high potential hazard

D.0.9 Classification of debris flow by scale may be determined in accordance with Table D.0.9.

Table D.0.9 Classification of debris flow by scale

Indicator	Type			
	Extremely Large	Large	Medium	Small
Total volume of one debris flow event (m^3)	≥ 500000	50000 (inclusive) to 500000	10000 (inclusive) to 50000	< 10000
Peak discharge of debris flow (m^3/s)	≥ 1000	200 (inclusive) to 1000	50 (inclusive) to 200	< 50

NOTES:
1 The debris flow caused by melting of snow and ice usually has a greater scale than that caused by rainstorm, the values in the above table may be raised based on actual situation.
2 The higher class shall be adopted when the results are in different classes by the two indicators.

D.0.10 Classification of debris flow by hazard may be determined in accordance with Table D.0.10.

Table D.0.10 Classification of debris flow by hazard

Indicator	Type			
	Extremely large	Large	Medium	Small
Death toll (person)	≥ 30	10 to 29	3 to 9	< 3
Direct economic losses (million CNY)	≥ 10	5 (inclusive) to 10	1 (inclusive) to 5	< 1

NOTE The higher class shall be adopted when the results are in different classes by the two indicators.

D.0.11 Classification of debris flow by potential risks may be determined in accordance with Table D.0.11.

Table D.0.11 Classification of debris flow by potential risks

Indicator	Type			
	Extremely large	Large	Medium	Small
Directly threatened person (person)	≥ 1000	500 to 999	100 to 499	< 100
Potential direct economic losses (million CNY)	≥ 100	50 (inclusive) to 100	10 (inclusive) to 50	< 10

NOTE The higher class shall be adopted when the results are in different classes by the two indicators.

D.0.12 Erosion degree classification of loose deposits shall be in accordance with Table D.0.12.

Table D.0.12 Erosion degree classification of loose deposits

Erosion degree	Identification indexes	
	Profile with intact soil layer	**Profile with less intact soil layer**
Extremely severe	Layers A and B are absent, and Layer C outcrops, subjected to denudation	Soil layers above Layer C are all denuded. And Layer C has been denuded partly
Severe	Layer A is absent, and Layer B outcrops, subjected to denudation	70 % of soils above Layer C has been denuded
Moderate	Residual thickness of Layer A is greater than 1/3, and Layers B and C are intact	30 % to 70 % of soils above Layer C has been denuded
Slight	Residual thickness of Layer A is greater than 1/2, and Layers B and C are intact	Less than 30 % of soils above Layer C has been denuded
Invisible	The profiles of the Layers A, B and C remain intact	The soils above Layer C remain intact

NOTES:

    1 In the table above, Layer A refers to topsoil, Layer B refers to subsoil, and Layer C refers to residual and weathered soil.

    2 "Profile with intact soil layer" indicates that soil profile is intact and can be divided into layers of A, B and C. "Profile with less intact soil layer" indicates that soil profile is less intact and mainly dominated by residual and weathered soil, which is difficult to distinguish the Layers A and B.

Appendix E Deformation Observation Items of Underground Cavern and Tunnel Surrounding Rock

Table E Deformation observation items of underground cavern and tunnel surrounding rock

Item	Content	Instrument	Purpose
Convergence deformation	The change in the spacing of wall surfaces and deformation rate	Convergence meter, displacement gauge	Analyze surrounding rock stability and support effect; estimate deformation magnitude based on the deformation rate
	Settlement in crown	Precise level, convergence meter, multipoint displacement meter	Analyze stability of crown and surrounding rock mass
	Floor upheaval	Precise level, convergence meter, multipoint displacement meter	Analyze necessity of rock bolts and concrete placement on inverted arch and the best time to anchor
Deformation inside the surrounding rock	Relative deformation between points on the wall and inside the surrounding rock	Multipoint displacement meter, sliding micrometer	Analyze loosened zone of surrounding rock, and reasonably determine the rock bolt length and the deformation and range inside the surrounding rock
	Horizontal and vertical displacement at a certain point inside the surrounding rock	Multipoint displacement meter, inclinometer, sliding micrometer	Analyze the rock mass condition before excavation, the stability of unexcavated rock mass at the excavation face, and the deformation distribution inside surrounding rock
Slippage of rock mass	Ground surface displacement and tilt displacement	Displacement gauge, joint gauge, inclinometer	Predict block sliding

Table E *(continued)*

Item	Content	Instrument	Purpose
Slippage of rock mass	Horizontal and vertical displacements in deep-seated rock	Multipoint displacement meter, TDR, inclinometer	Analyze sliding surface position and direction
Rotation of rock mass	Angular displacement, tilt	Inclinometer	Analyze angular displacement and tilt change of rock mass
Loosening range of surrounding rock	Measurement of velocity and amplitude of sound wave and seismic wave	Acoustical detector, seismograph	Analyze loosening range of surrounding rocks because of rock burst, and stress release, design rock bolt and other supports
State of ground surface and structure	Ground subsidence and upheaval	Level, settlement gauge	Analyze the range affected by tunnel excavation and the stability of rock mass above the tunnel
State of ground surface and structure	Structure subsidence, upheaval and tilt	Level, settlement gauge	Analyze the influenced range and safety of structures

Explanation of Wording in this Specification

1. Words used for different degrees of strictness are explained as follows in order to mark the differences in executing the requirements in this specification:

    1) Words denoting a very strict or mandatory requirement:

        "Must" is used for affirmation, "must not" for negation.

    2) Words denoting a strict requirement under normal conditions:

        "Shall" is used for affirmation, "shall not" for negation.

    3) Words denoting a permission of a slight choice or an indication of the most suitable choice when conditions permit:

        "Should" is used for affirmation, "should not" for negation.

    4) "May" is used to express the option available, sometimes with the conditional permit.

2. "Shall meet the requirements of…" or "shall comply with…" is used in this specification to indicate that it is necessary to comply with the requirements stipulated in other relative standards and codes.

List of Quoted Standards

GB/T 12897,	Specifications for the First and Second Order Leveling
GB/T 18314,	Specifications for Global Positioning System (GPS) Surveys
GB/T 19531.1,	Technical Requirement for the Observational Environment of Seismic Stations—Part 1: Seismometry
DL/T 5178,	Technical Specification for Concrete Dam Safety Monitoring
DL/T 5416,	Specification of Strong Motion Safety Monitoring for Hydraulic Structures
DB/T 16,	Specification for the Construction of Seismic Station Seismograph Station
DB/T 47,	The Method of Earthquake-Related Crust Monitoring—Fault-Crossing Displacement Measurement